New Age Fibres for High Performance Applications

R.Senthil Kumar,

Head – Yarn R&D,

SSIL- Textile Division,

sen29iit@yahoo.co.in

Content

- High performance Fibres
- Fibre property requirement for High performance applications
- Specific Strength
- Energy Absorbing capacity
- Commodity Fibres
- Typical properties of High Performance Fibres
- Aramid Fibres
  - Manufacturing
  - Types & Forms
  - Molecular Structure
  - Characteristics
  - Kevlar - types and its properties
- Carbon Fibres
  - Manufacturing
  - Forms
  - Advantages and Disadvantages
  - Applications

Content

- **Ceramic Fibres**
  - Types
  - Manufacturing process
- **Glass Fibres**
  - Classification
  - Manufacturing process
  - Chemical composition
  - Advantages and Disadvantages
  - Applications
- **PBO fibre**
  - Manufacturing
  - Properties
  - Applications
- **Polyethylene**
  - Manufacturing
  - Properties
  - Applications

High Performance Fibres

- In a sense, all fibres except the cheapest commodity fibres are high performance fibres.

- The natural fibres (cotton, wool, silk . . .) have a high aesthetic appeal in fashion fabrics.

- With the advent of manufactured fibres (rayon, acetate, nylon, polyester . . .) in the first half of the twentieth century, not only were new high-performance qualities available for fashion fabrics, but they also offered superior technical properties.

- A replacement of natural and regenerated fibres by synthetic fibres occurred in most technical textiles.

Conventional fibres	High performance and Speciality fibres
Volume Driven	Technically Driven
Price Oriented	Application Oriented
Large Scale Production	Smaller Batch Production

High Performance Fibres

- The combination of moderately high strength and moderately high extension gives a very high energy to break, or work of rupture.

- Good recovery properties mean that they can stand repeated high-energy shocks.

- The high-stretch characteristics of elastomeric fibres, such as Lycra, have an undeveloped potential for specialized technical applications.

Important terms

- Strength
- Tensile Strength
- Compressional Strength
- Tenacity/ Specific strength
- Toughness
- Flexural Strength
- Torsional Strength
- Impact Strength
- Elongation
- Modulus – **material's resistance to deformation**

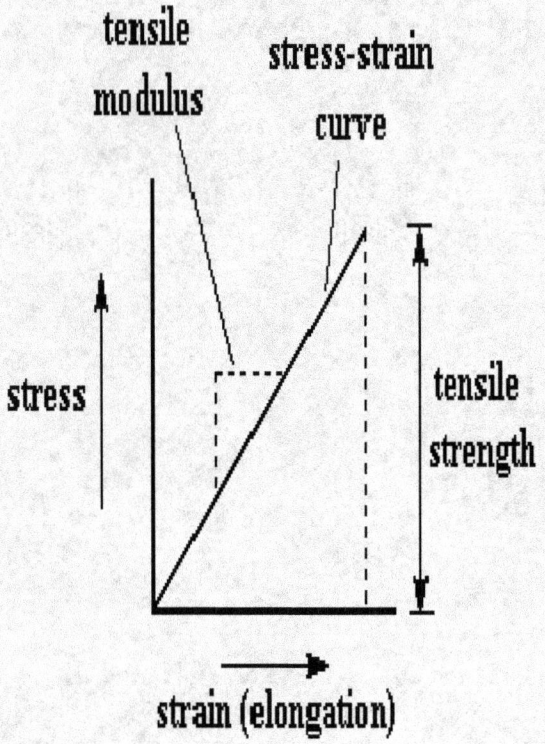

tensile modulus

stress-strain curve

stress

tensile strength

strain (elongation)

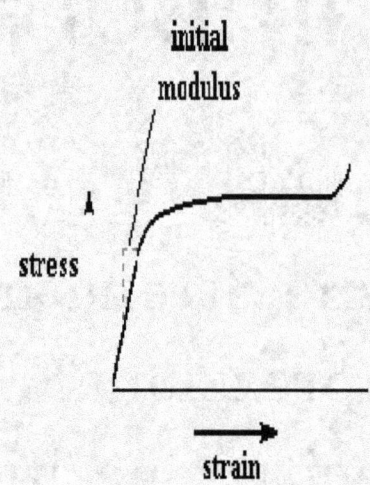

initial modulus

stress

strain

After Odian, George; *Principles of Polymerization,* *3rd ed.,* J. Wiley, New York, 1991, p.34.

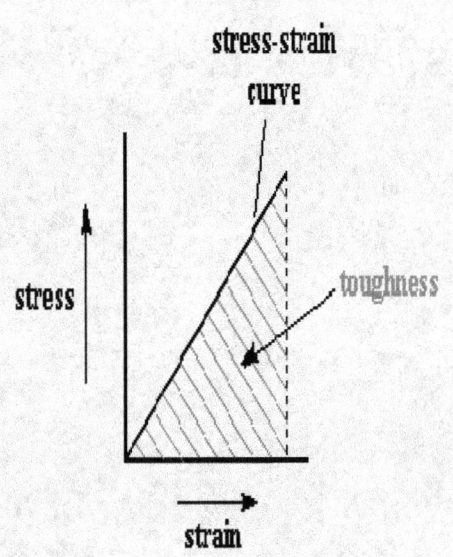

stress-strain curve

stress

toughness

strain

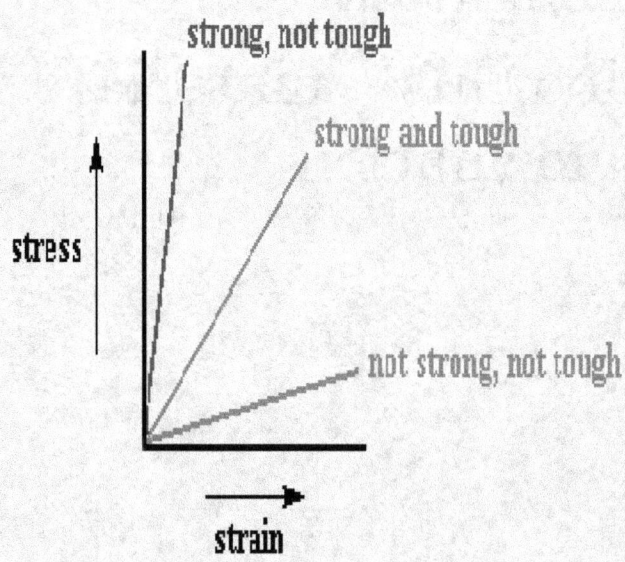

strong, not tough

strong and tough

stress

not strong, not tough

strain

Mechanical Properties of Fibre forming Polymers

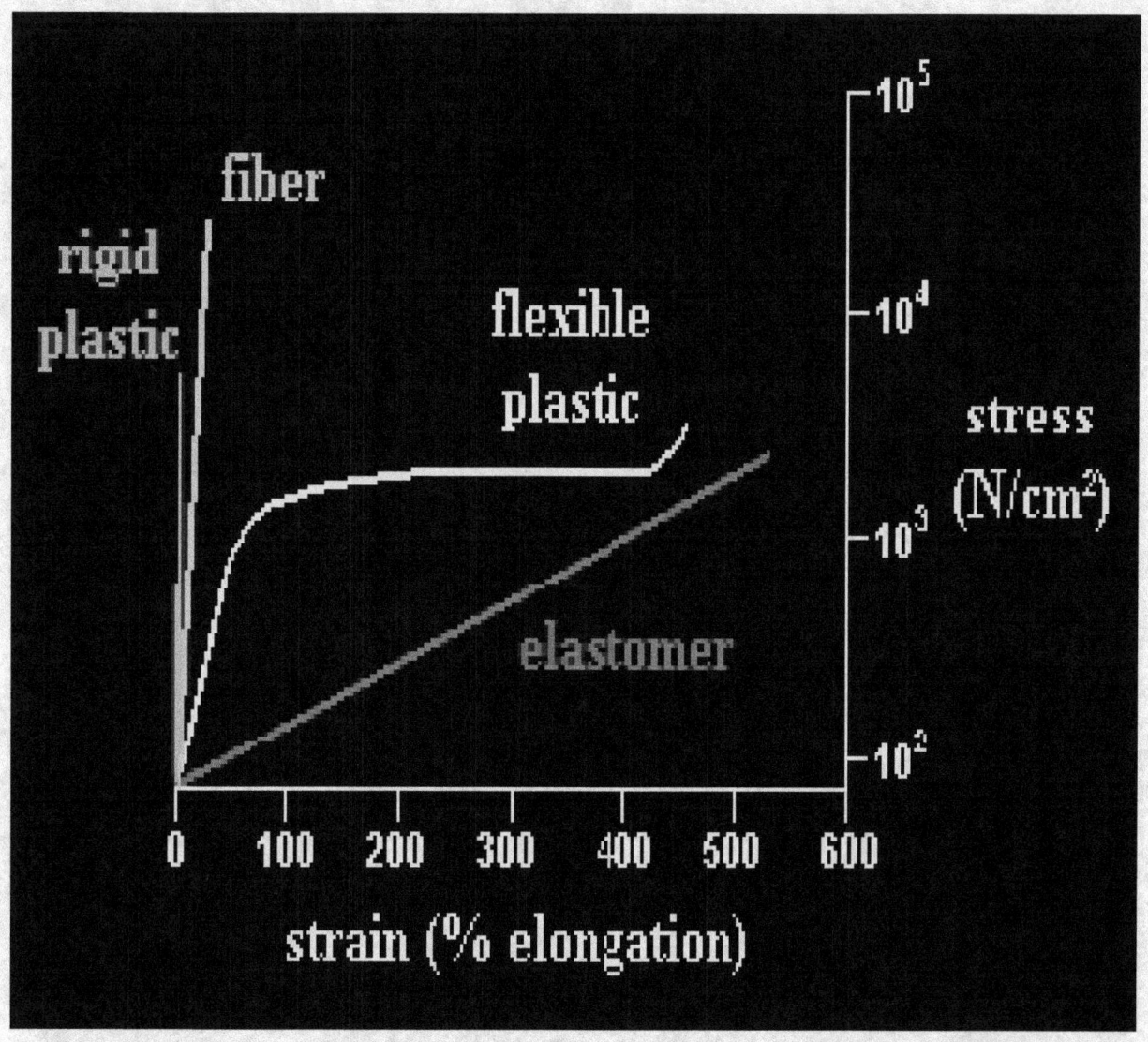

After Odian, George; *Principles of Polymerization, 3rd ed.*, J. Wiley, New York, 1991, p.34.

Fibre Property Requirement for Different Applications

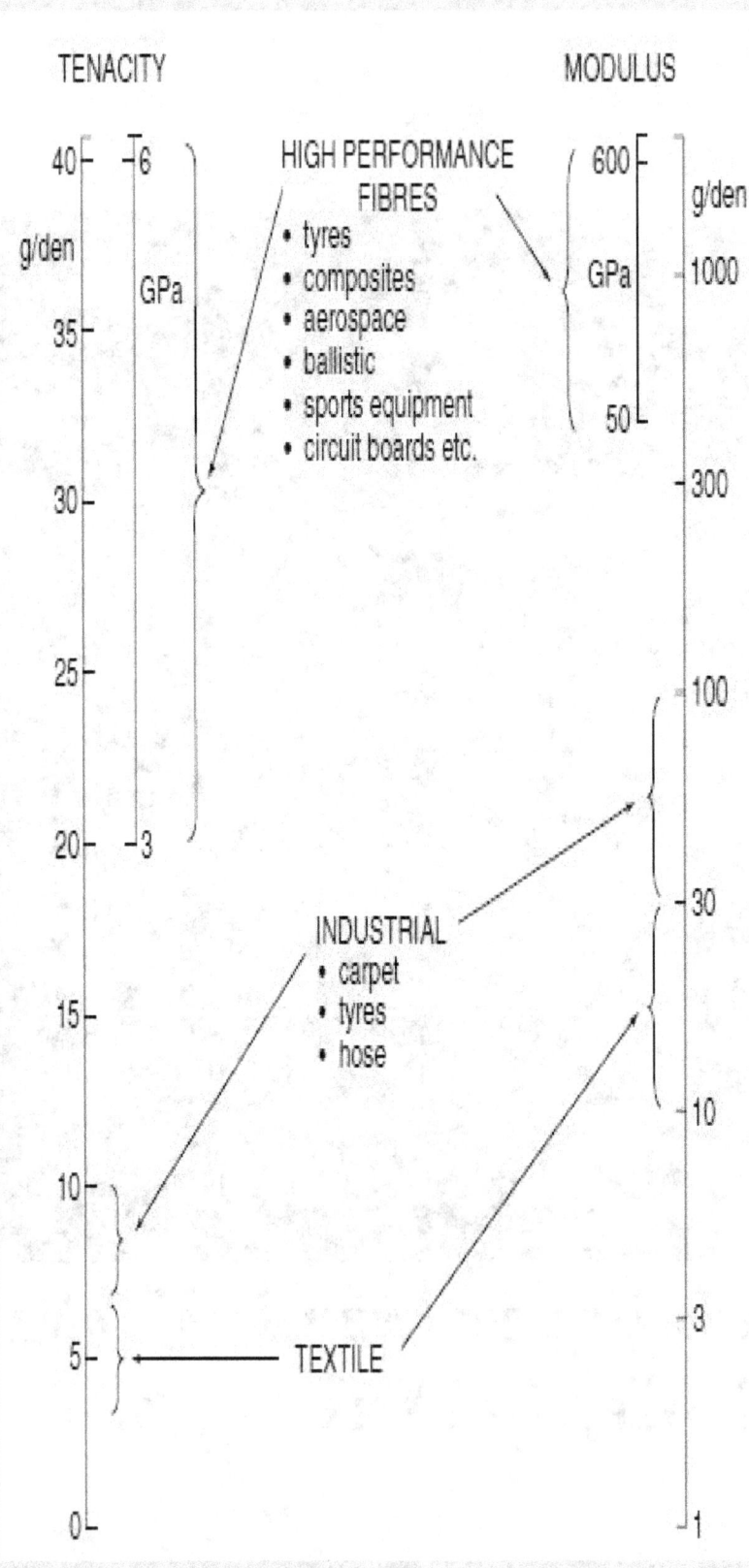

Specific Strength

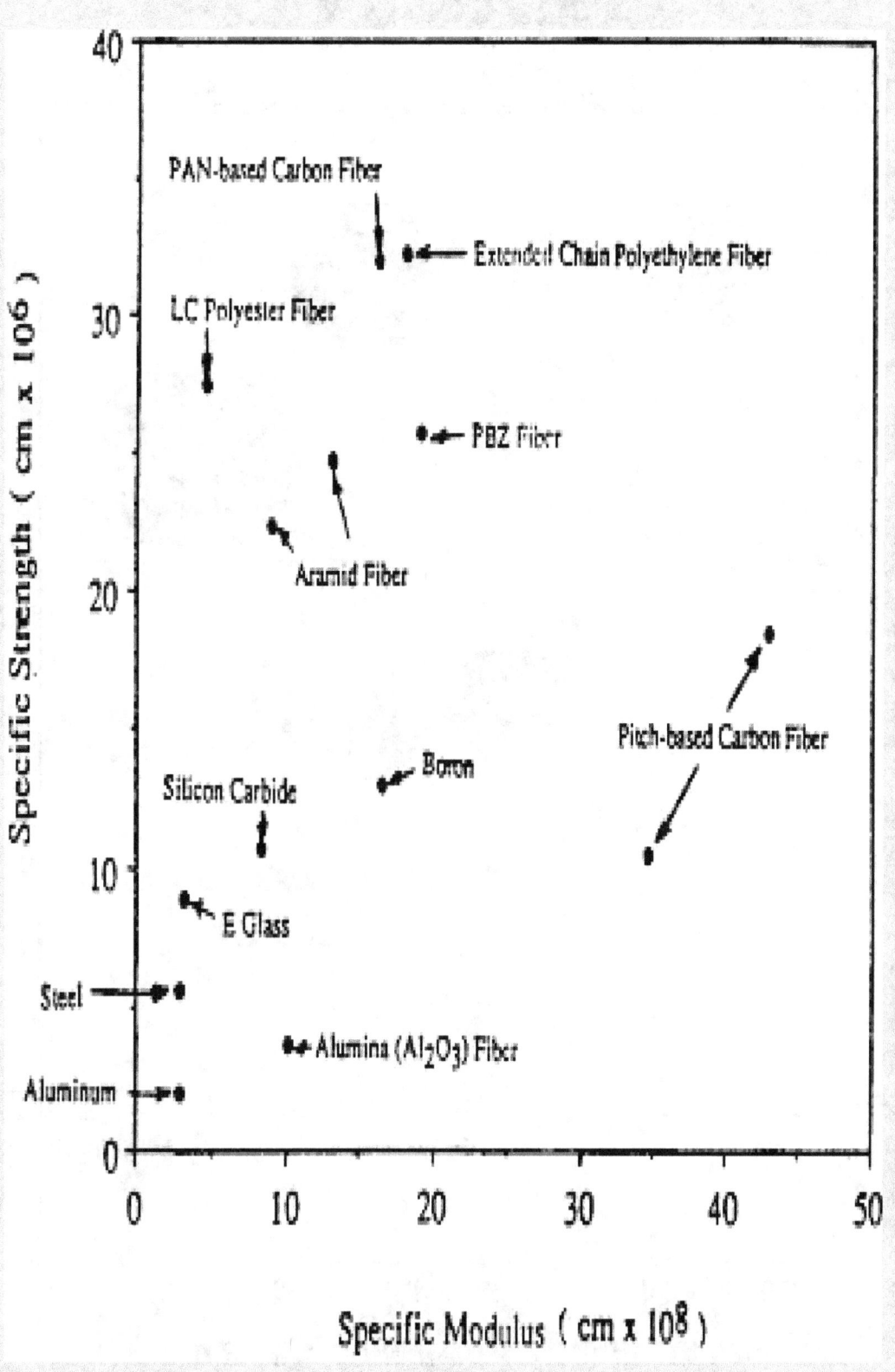

Energy Absorption Capacity

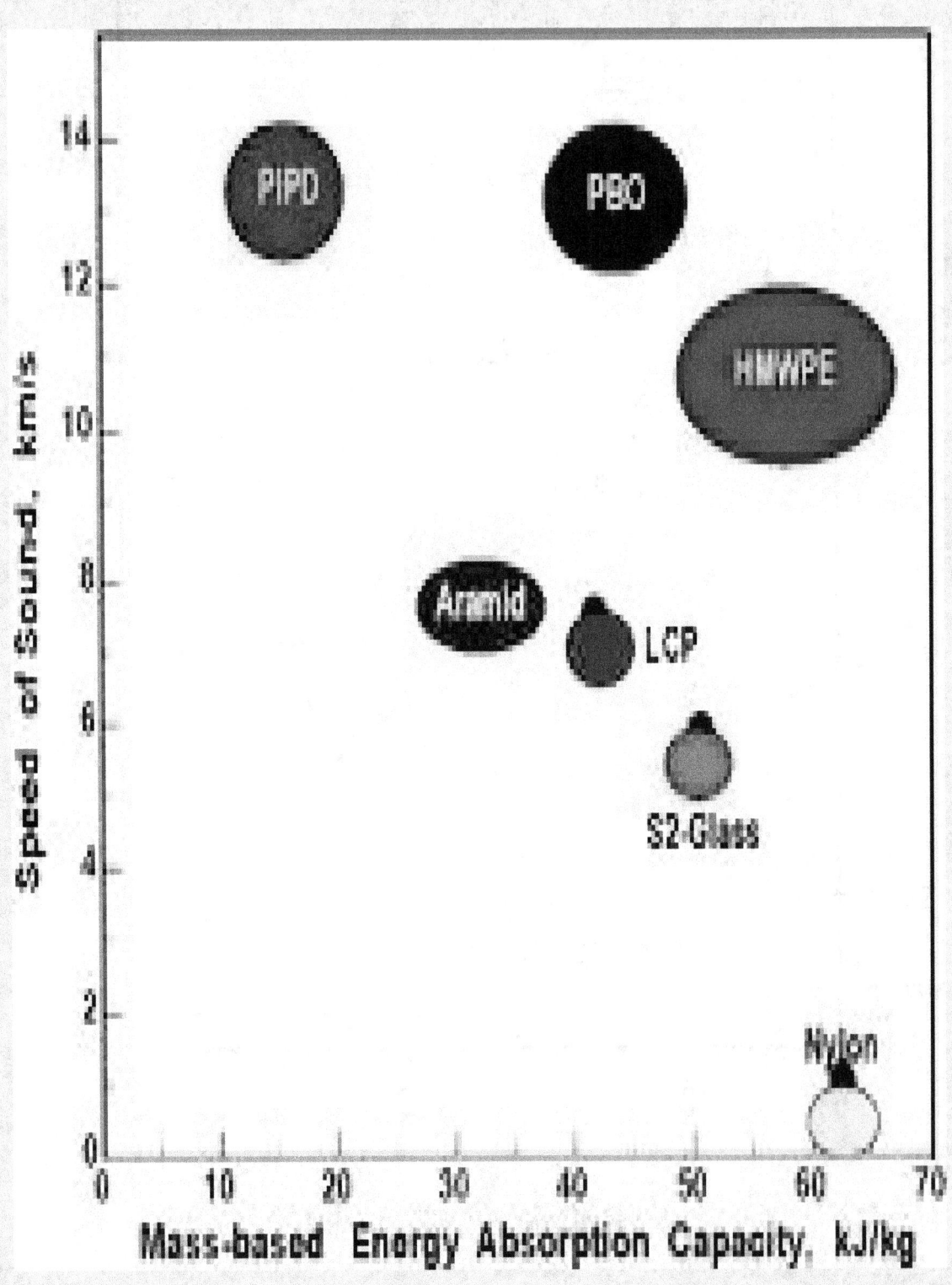

Commodity fibres

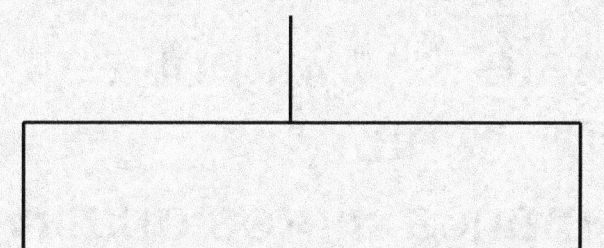

Natural : cotton, wool, silk   Manmade/manufact
ured

Natural :

100 years ago—natural fibres were used in engineering
applications also~ technical/industrial textiles

e.g.: reinforced in tyres

cotton~1900

improved rayons~ 1935 to 1955

nylon polyester& steel~ dominate the market totally

Aesthetic appeal in fashion fabrics :Clothing ,Upholstery ,
Carpets

Manmade (1st Generation Fibers)

Rayon, acetate, Nylon, polyester
1st half of 20th century
High performance fibres quality for: fashion fabrics

Superior Technical properties

Commercial nylon and polyester fibres~ 10g/den

$$(\sim 1N/tex) \text{ or}$$

1GPa

Extension at break~

>10%

High performance fibres ~ High strength (tenacity)

High modulus

Inorganic Fibres Fibers

Glass

Carbon(quasi organic)

HPPE

Ceramic
from linear

PBO

polyesters

Polymeric

Aramids

Gel spun

HM-HT fibres

Polymer e.g.

aromatic

2nd generation fibres :Thermal or chemical resistant fibres.

3rd generation fibres : **s**mart fibres (special physical or chemical properties)

4th generation fibers : Responsive to environment e.g. Temp, pH etc.

e.g. Soft swatch~ fibres that become conductive under pressure

Fibres	Strength	Modulus
Cotton, wool , silk ~ flax/ramie (higher)	0.1-0.4N/tex	2 to 5 n/tex
Regenerated cellulose (acetate/viscose rayon)	<0.2N/tex(earliest) 0.4N/tex 0.6N/tex	
Tenasco. (tyre cords) 1960	0.6 N/tex	16N/tex
Rostisan 1945	1.1 N/tex	30N/tex
Bocell (azko/nobel)	0.5N/tex 0.5N/tex	
Nylon 1938	0.8N/tex	2.5N/tex 10N/tex
PET	2N/tex	
Nylon/PET tyre	3.5N/tex	

Fibre	Strength	Modulus
High Strength Carbon fibres	3N/tex 5Gpa	800Gpa 400N/tex
Glass fibres	1.6N/tex (4Gpa)	35N/tex (90Gpa)
Ceramic fibres	1N/tex (3GPa)	100N/tex (400GPa)

TYPICAL PROPERTIES

FIBER	STRENGTH (KSI)	MODULUS (MSI)	STRAIN (%)
E-Glass	350	6	2
S-Glass	500	6	3
CF-Pan	600	33-50	2
C-Pitch GP	200	6	0.3
Pitch UHM	400	70-120	0.5
Aramid	500	10-20	2
Ceramic	100	10-40	2
Nylon	50	0.5	5-50

FIBER PROPERTIES

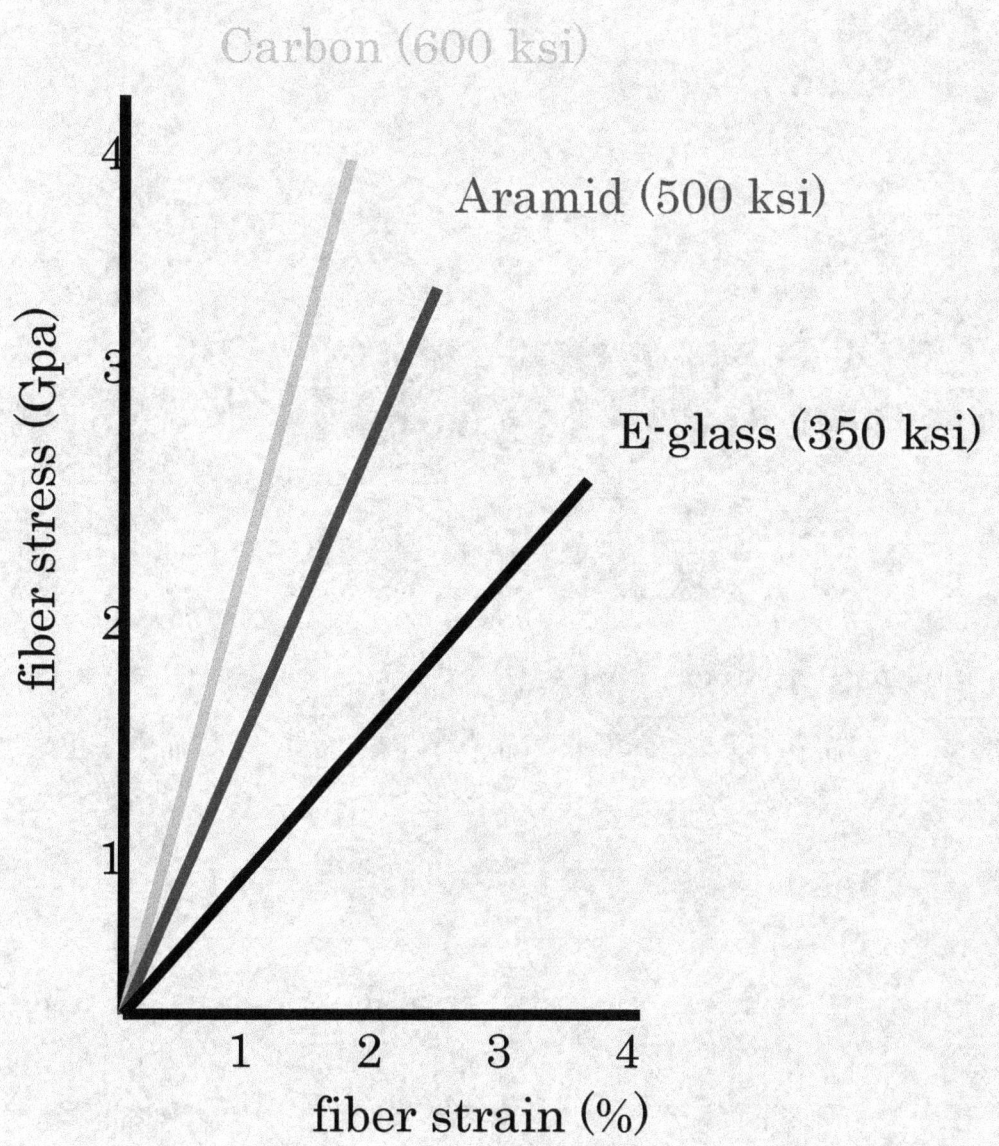

FIBER PROPERTIES - TENSILE STRENGTH

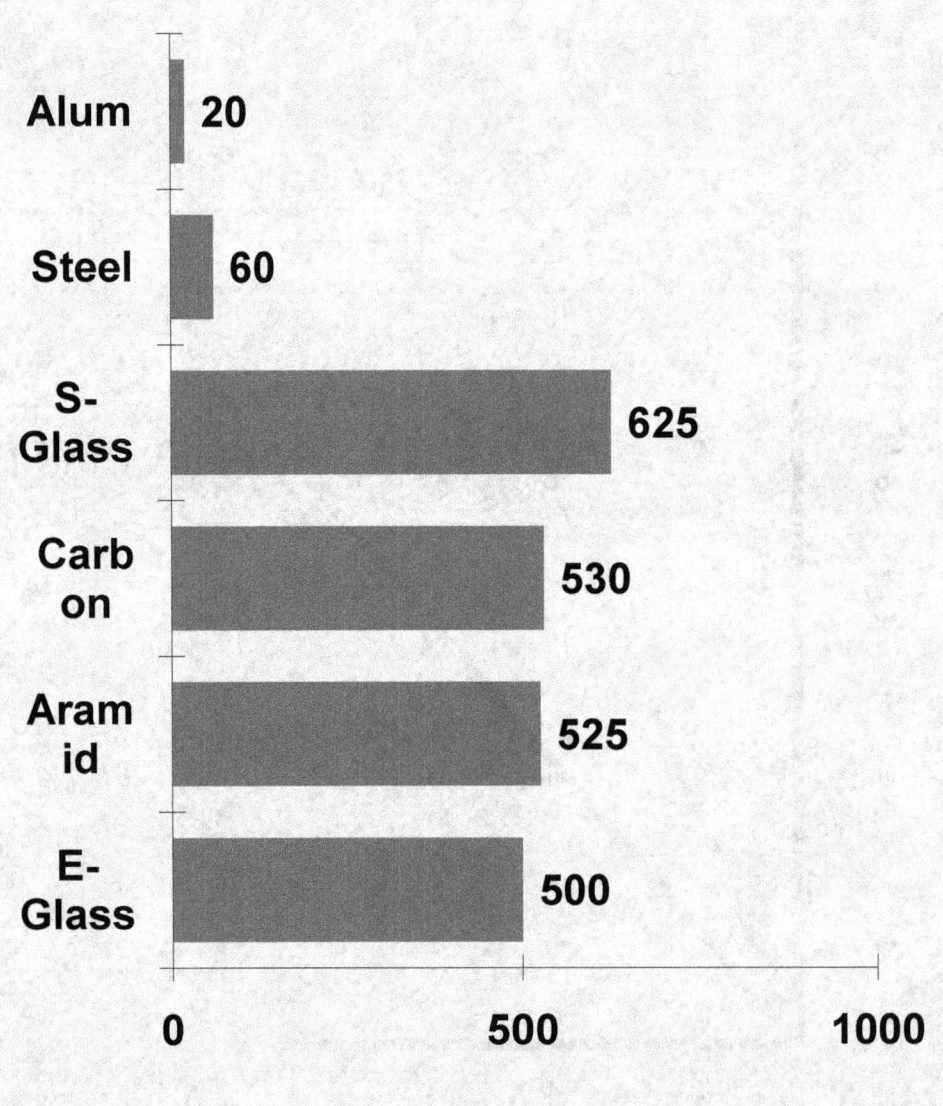

x10^3 psi

FIBER PROPERTIES - STRAIN TO FAILURE

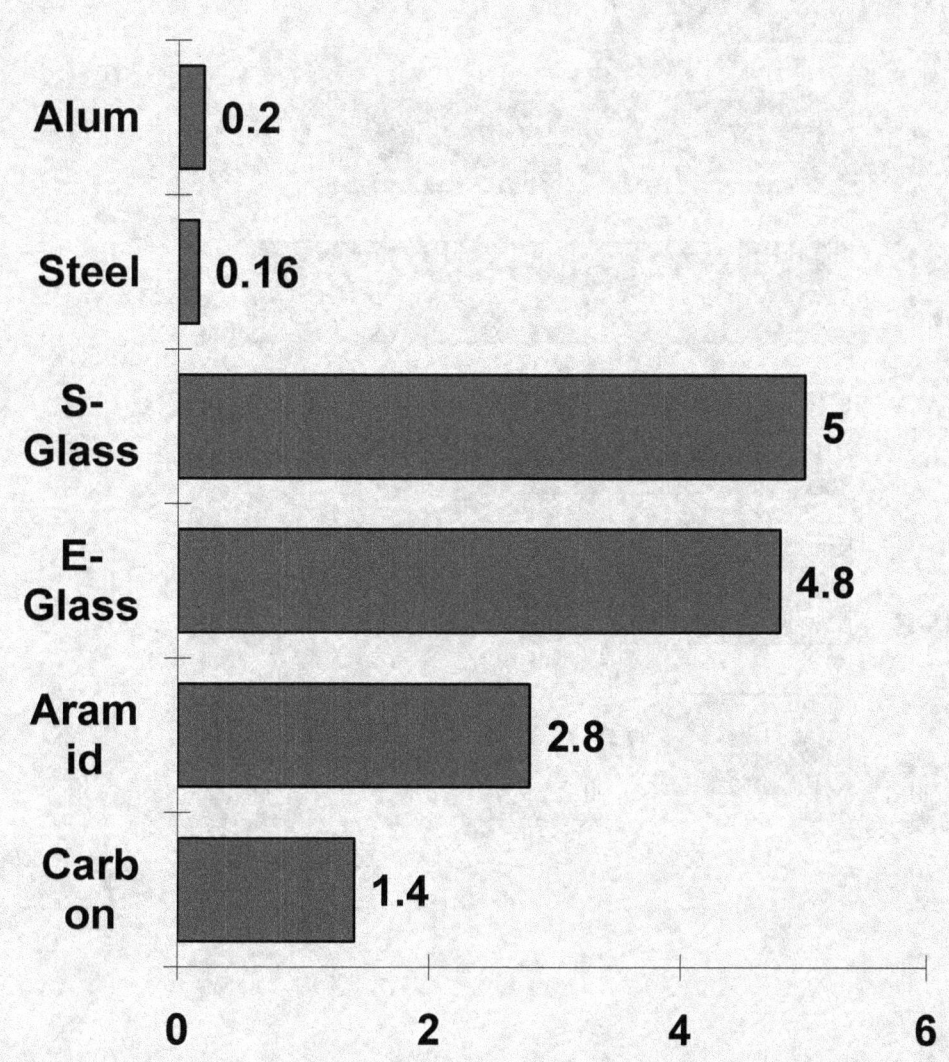

(%)

FIBER PROPERTIES - TENSILE MODULUS

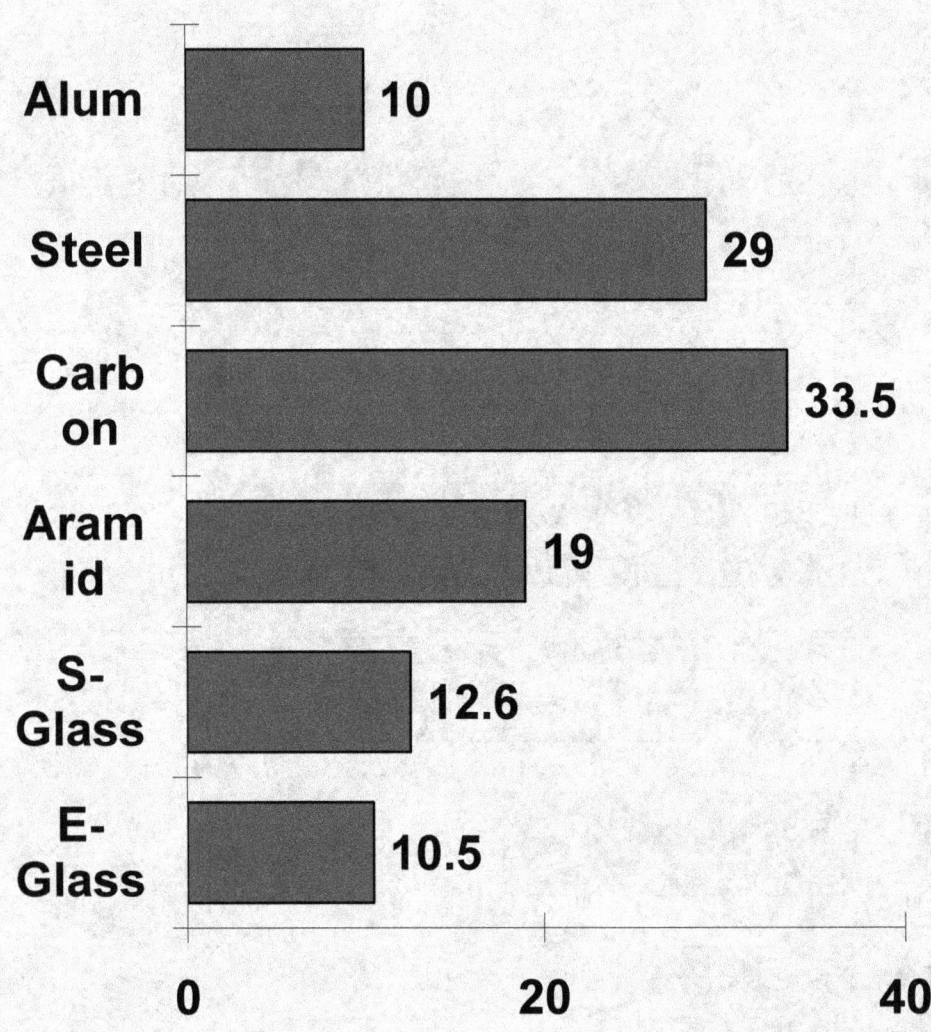

10⁶ psi

FIBER PROPERTIES - CTE -
Longitudinal

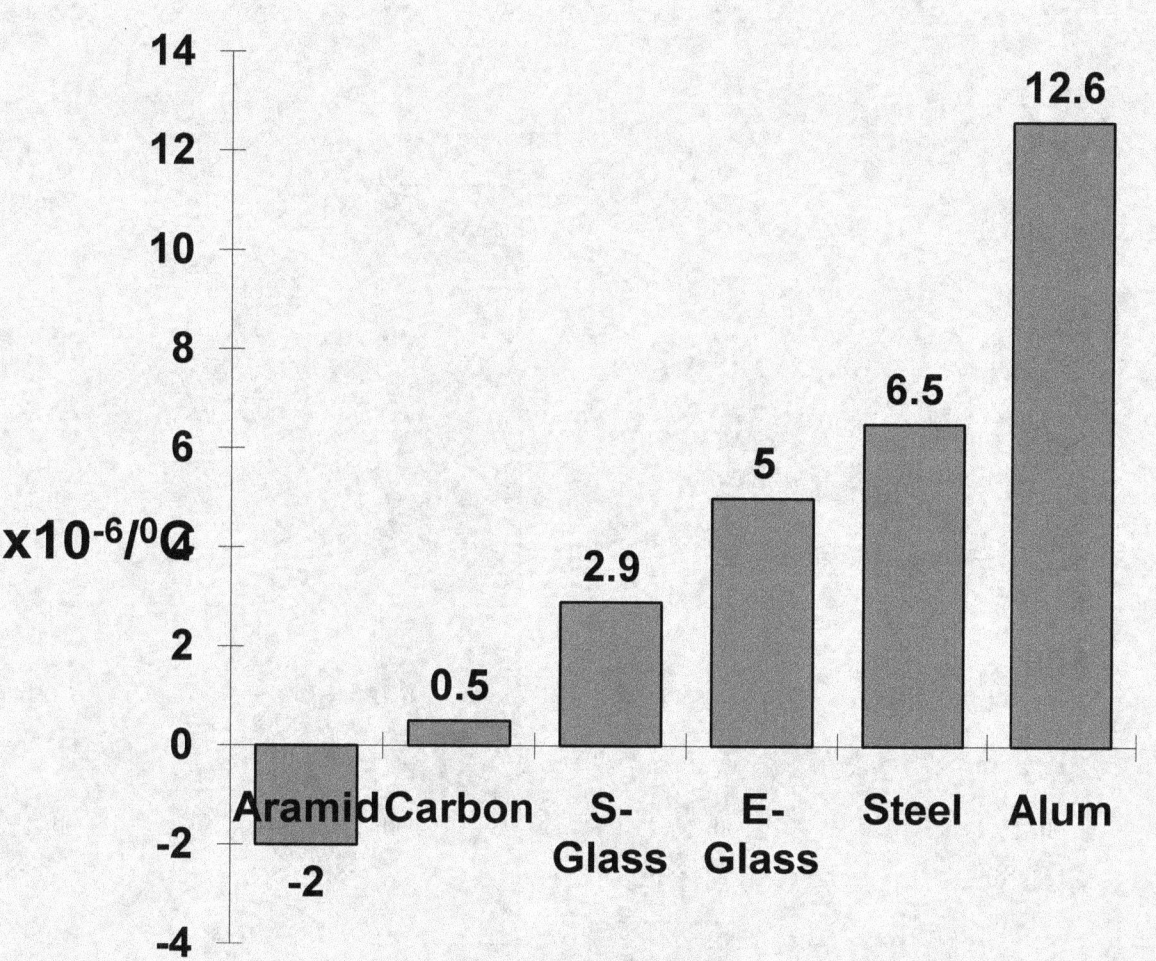

FIBER PROPERTIES - THERMAL CONDUCTIVITY

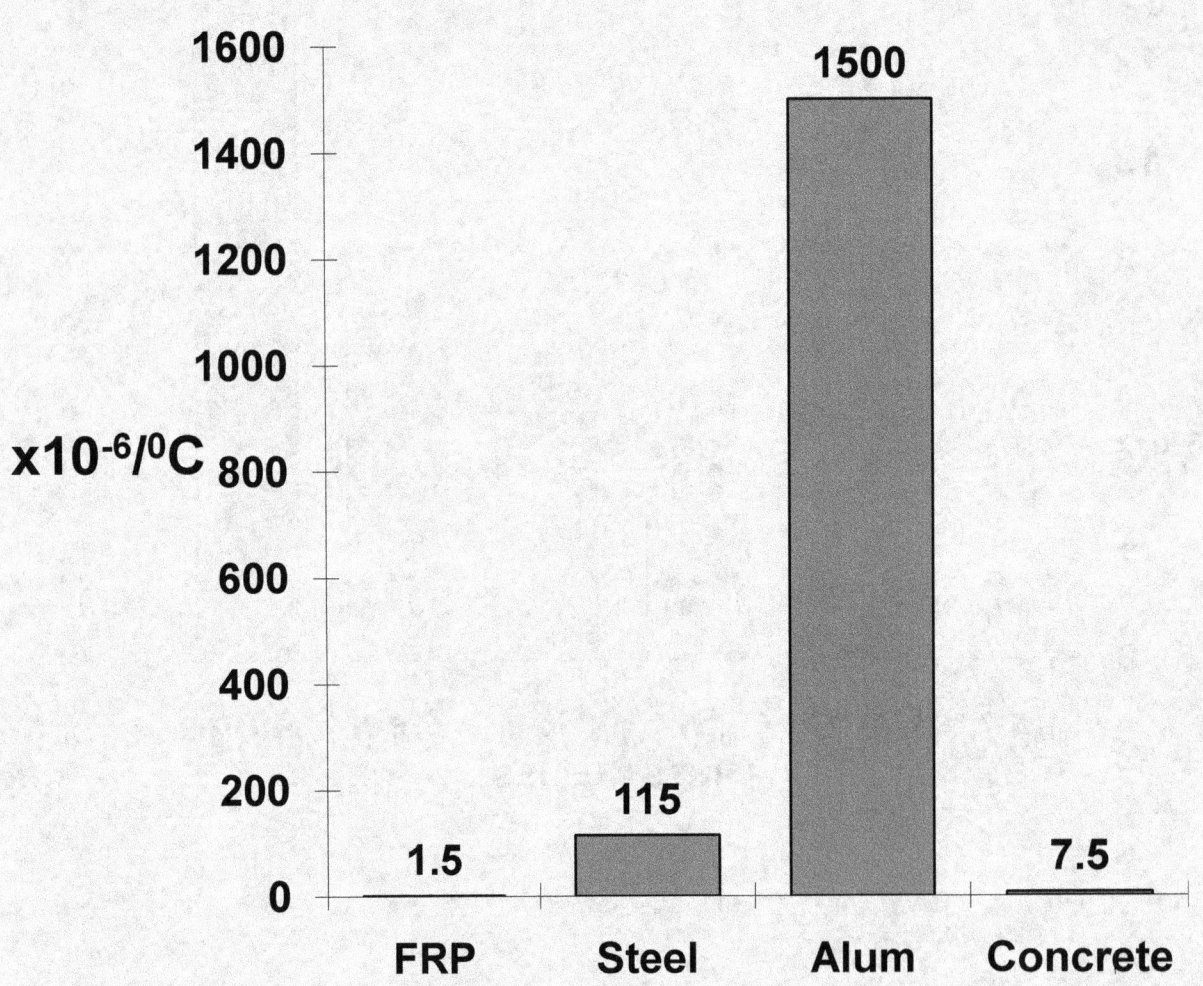

BTU-in/hr-ft^2 - ^{0}F

ARAMID
FIBRES

Kevlar (synonymous) with Aramids

✓Dupont in 1971

✓Organic fibres (aromatic polyamide)

Manufacture:

Simple regular polymeric structure, rings in the backbone, LCP, solution spinning, dissolved in H_2SO_4, extruded 80°C, dry-jet-wet, narrow air-gap, 6 m/s speed, quenched at 1°C water, washed, dried, wound, not brittle.

• LCP allows orientation – solidification, heat treatment under tension, longitudinal modulus, high x, rigid molecule – high temperature stability.

Types and Forms :

✓ Diameter ~ 12 µm, yarns with couple of dozens to several thousand
✓ Kevlar 29, 49, 149 (high toughness, high, ultrahigh Modulus)

• Untwisted yarns (100 – 1000 filaments per yarn)
• Roving (3000 – 5000 filaments per roving)
• Fabrics

Molecular Structure of Kevlar – 49 fibre

✓ Aromatic polyamide (Aramids)
✓ $T_g \sim 360\ ^0$ C
✓ Tm $\sim 560\ ^0$C

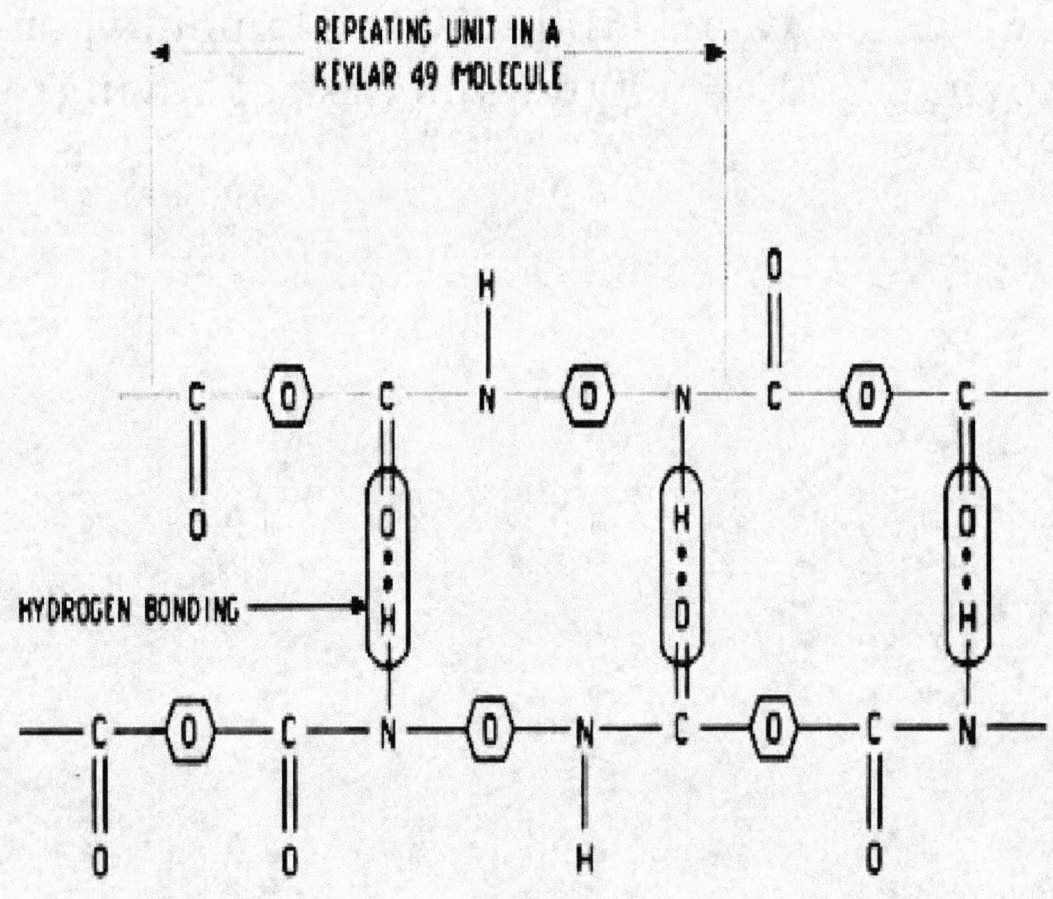

Characteristics

- Partly crystalline, High orientation, Light weight (lowest specific gravity) ($\rho \sim 1.45$), Filament is highly anisotropic. Better properties in longitudinal direction.

- Highest tensile strength to weight ratio amongst current reinforcing fibres (3.0 GPa); compressive strength 1/100 to longitudinal strength, Negative coefficient of thermal expansion in longitudinal direction

- High modulus (aromatic ring) (130GPa), Better chemical and thermal stability. Over other organic fibres (except few strong strong acids and alkalis) Max long term use temperature $\sim 160°$ C .

- Kevlar does not melt or support combustion upto $427°$ C, but will start decomposition beyond this temperature. Excellent toughness – difficult to cut . Special tools for machining, moisture sensitivity, damaged by kink band formation.

KEVLAR - Types

Type	Tenacity (mN/tex)	Initial modulus (N/tex)	Elongation at break (%)
Kevlar® 29	2030	49	3.6
Kevlar® 49	2080	78	2.4
Kevlar® 149	1680	115	1.3
Nomex®	485	7.5	35
Twaron®	2100	60	3.6
Twaron® High Modulus	2100	75	2.5
Technora®	2200	50	4.4

Application :

✓ Bullet proof vests (non-composite), Aramid has negative CTE,

✓ Zero thermal expansion of composites, applications limited due to moisture sensitivity;

✓ Leading edges of wings in aircraft, Ballistic Armor, Sporting- goods, Break lining, Marine, Soft, light weight body armours & helmets. (protecting police officers & military personnel)

Draw Backs

- ✓ Low compressive strength

- ✓ Low thermal conductivity

- ✓ High vibration damping coefficient

- ✓ Sensitive to UV light (matrix protects in a composite)

- ✓ Kevlar – 49 are hygroscopic (absolute moisture 6 % at 100% RH and 23° C)

- ✓ Little effect on tensile strength, tend to crack internally at the preexisting micro-voids and produce longitudinal splitting

Carbon Fibres

Manufacture :

✓Precursors – Rayon, PAN, (Melt) Pitch;

✓Fibres are drawn and oxidized at < 400°C to crosslink, carbonized at 800°C, pyrolysis [(-) O_2] to remove non-carbon elements

✓Graphitization at > 1000°C – remove impurities and increase crystallinity, drawing → Orient graphite layers in fibre direction, defects, surface treated and sized

✓Active nitrile group to produce ladder polymer with six-membered rings(in tension) in O_2 environment, further heating – graphite like structure skin-core,last temperature critical, high strength for PAN at 1500°C, strength ↑ T for pitch.

Thomas Alva Edison (1878) made carbon fibres from cotton/bamboo for filaments of lamp.

Courtaulds – 1950's carbon fibre for high temperature missile application. Highest strength and stiffness-versions of carbon fibres.

Advantages:
- ✓ Exceptionally high tensile strength – weight ratios as well as tensile modulus – weight ratios
- ✓ Tensile modulus is ~ 207 Gpa – 1035 Gpa
- ✓ Very low coefficient of thermal expansion
- ✓ High fatigue strength

Disadvantages:
- ✓ High cost
- ✓ High Electrical conductivity ('Shorting' in unprotected machinery)

Arrangement of carbon atoms in a graphite crystal

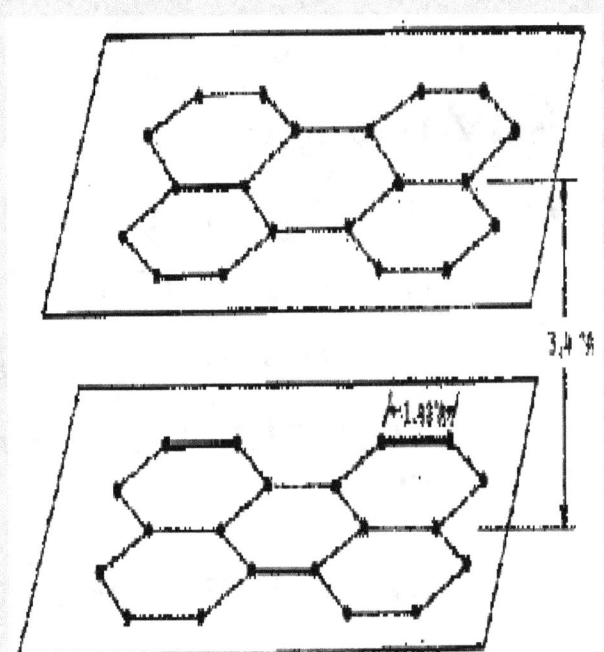

✓ Blend of Amorphous
&
Graphitic carbon

✓ Carbon atoms arranged
in planes of regular
hexagon

✓ Strong covalent
bonds between 'C'
atoms weaker
Vander Waals forces
between planes

✓ Highly anisotropic
physical and
mechanical
properties.

(a) Circumferential
(b) Radial
(c) Random
(d) Radial-circumferential
(e) Random-circumferential

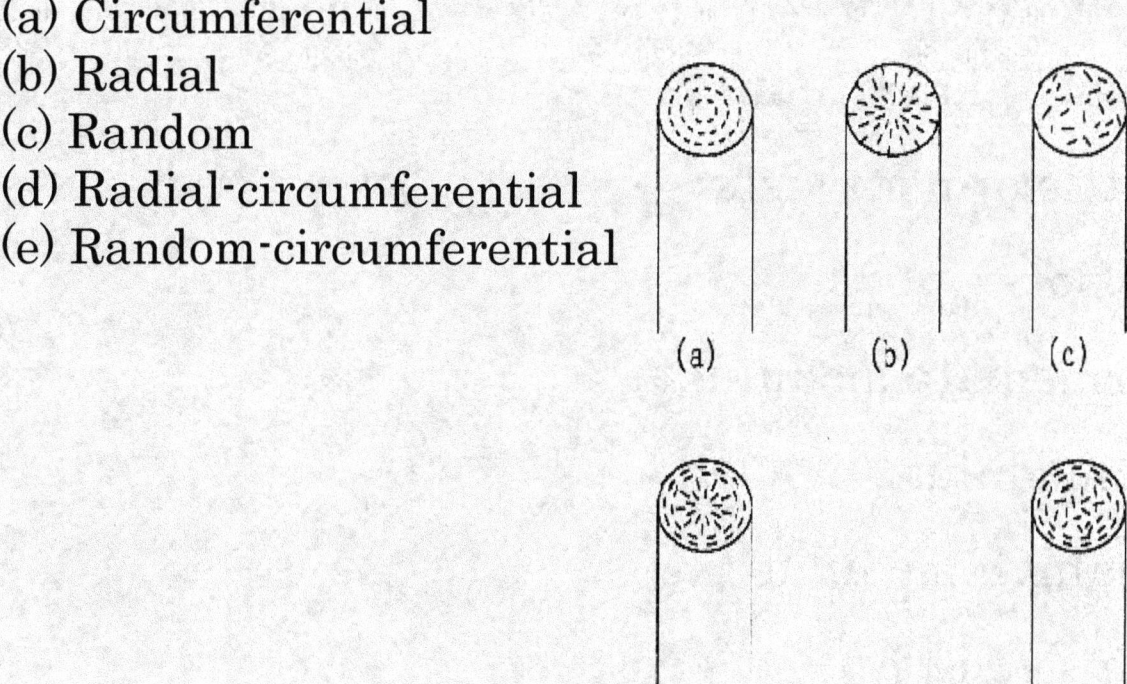

Commercial fibres – Circumferential arrangement in skin and radial or random arrangement in core

Types and Forms :

✓ 3K (3000 fibers), 6 K, 12 K,

✓ 4 – 11 μm (~ 1 μm).

✓ Little or not twist,

Version –

• intermediate modulus.

• low modulus.

• ultrahigh modulus.

• high strength

Three dimensional layers of graphene layers and disordered arrangement.

Types of carbon fibers

High Strength			High Modulus			Ultra High Modulus		
σ (Gpa)	E (Gpa)	ρ (g/cc)	σ (Gpa)	E (Gpa)	ρ g/cc	σ (Gpa)	E (Gpa)	ρ g/cc
3.65	231	1.762	5.65	290	1.81	1.52	483	1.96

Commercial forms

✓ Long & continuous tow (bundle of 1000 to 160,000 parallel filaments)

✓ Chopped (6 – 50 mm)

✓ Milled (30 – 3000 μm long)

✓ Woven (2-D fabrics of different weave styles.

Carbon fibre manufacturing (Flow diag.)

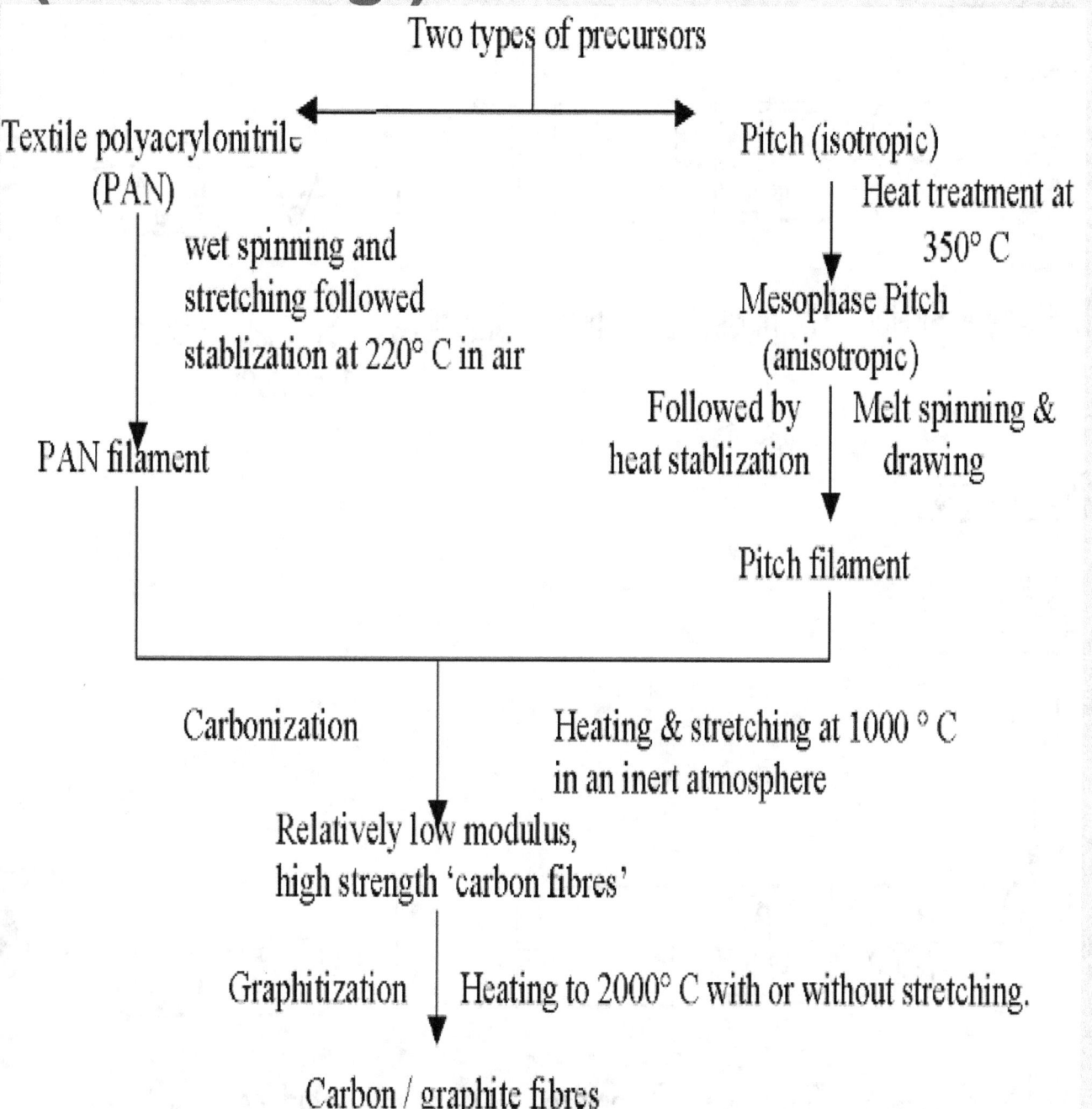

PAN process

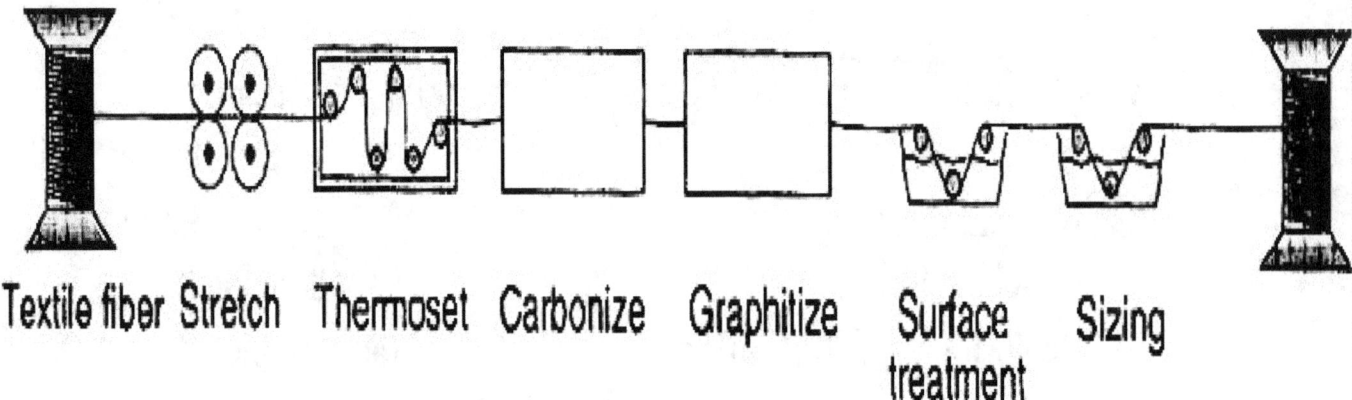

Textile fiber   Stretch   Thermoset   Carbonize   Graphitize   Surface treatment   Sizing

Pitch process

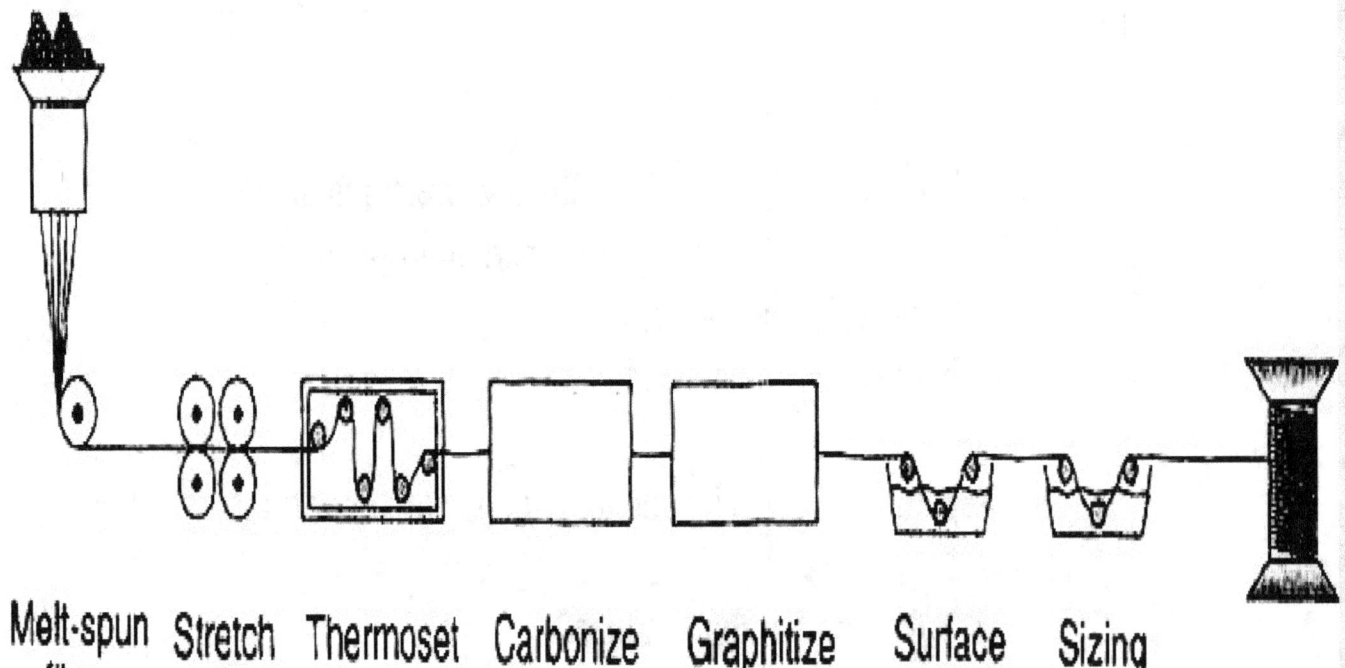

Melt-spun fiber   Stretch   Thermoset   Carbonize   Graphitize   Surface treatment   Sizing

Schematic of carbon fiber production. Redrawn from reference [15]

Applications

✓ Sports, fishing rods, golf clubs, rackets, filament wound rocket motor cases, pressure vessels, aircraft structural components, fixed wings, helicopter wings, body, stabilizer, rudder components.

✓ Polymorphs of carbon, diamond, hard, high thermal conductivity, transparent, films on other materials; graphite –hexagonal carbon atoms , planer structure.

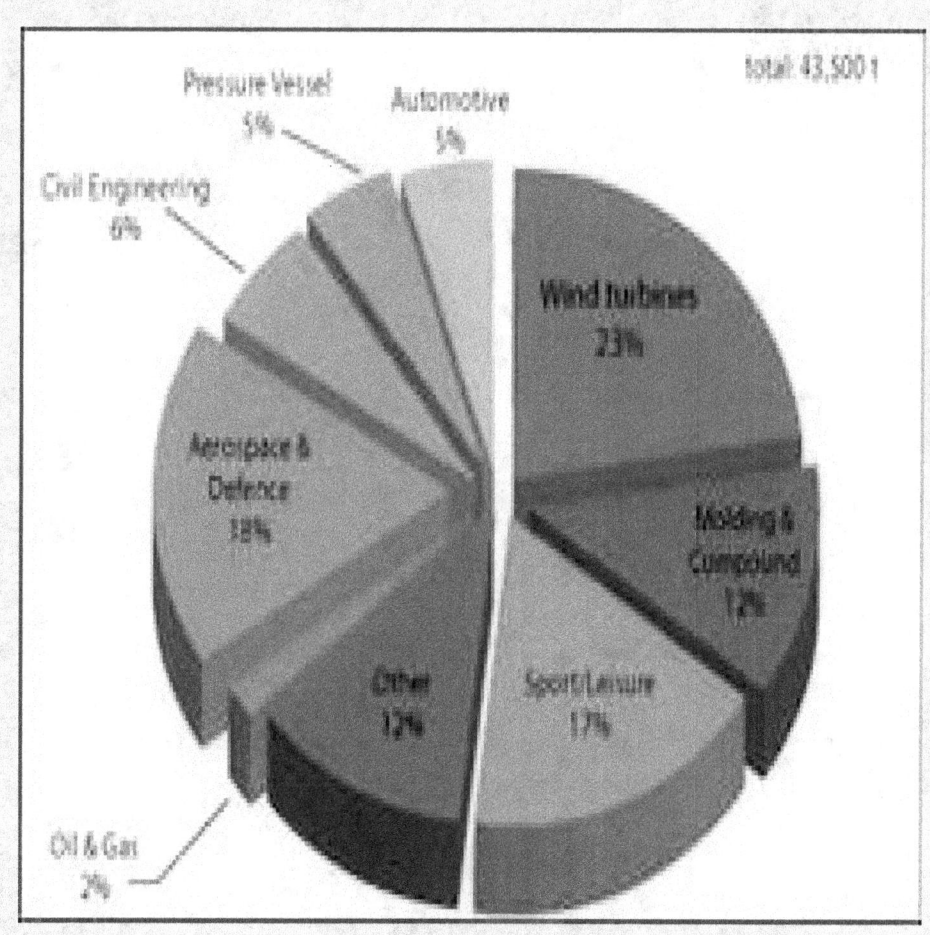

APPLICATION OF CFRP IN AIRCRAFTS

- A large-scale structure such as an aircraft wing can be fabricated with CFRP (integral molding) as a single component ,this may help sometimes to reduce weight by removing certain additional components sometimes.

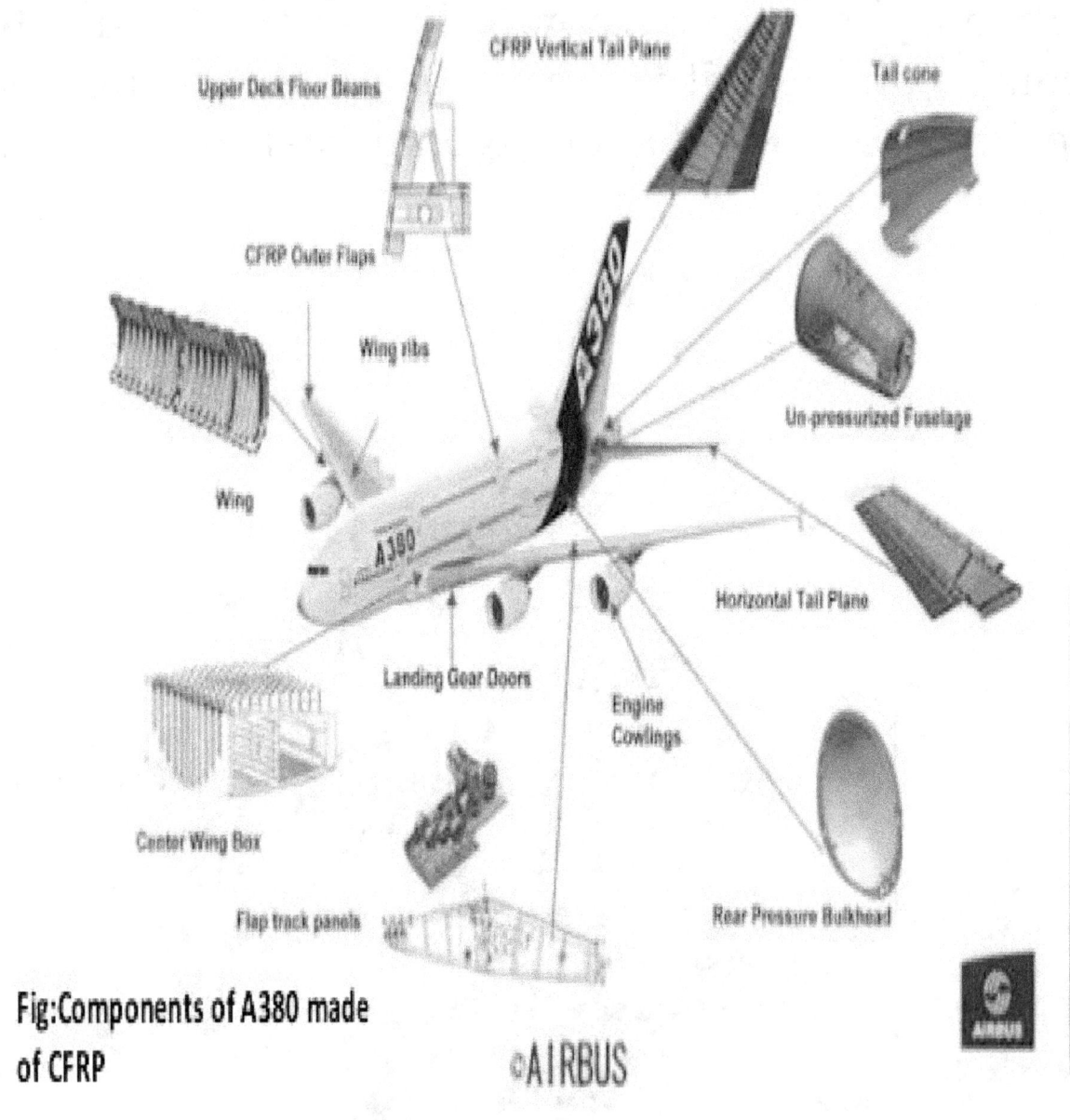

Fig:Components of A380 made
of CFRP

©AIRBUS

Lamborgini *"Sesto Elemento"* 2011

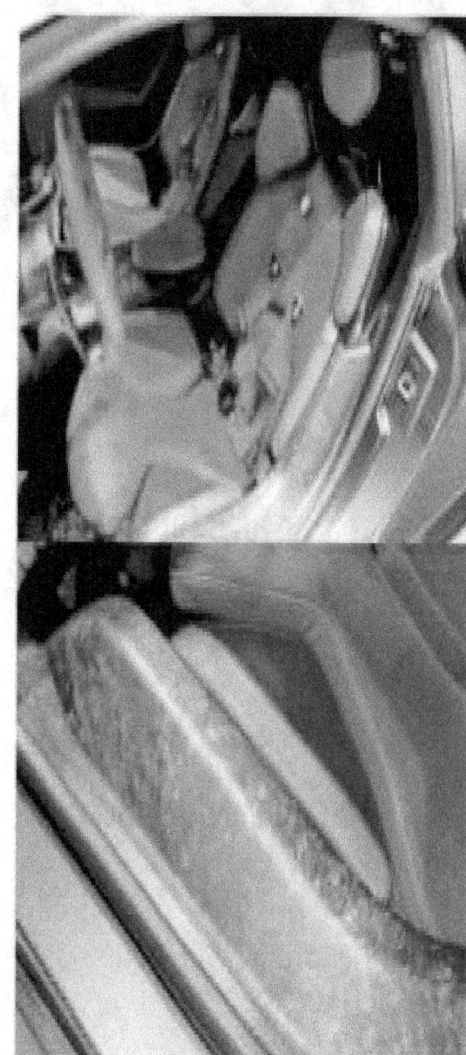

Lamborghini's **Sesto Elemento** - was a technology demonstrator :

- 80 % of the car is CRPF
- Featuring a skin one-third the thickness of previous CFRP sports car body panels, the car's monocoque achieves the required rigidity via integrated stiffeners
- Uses one-shot <u>Forged Composites</u> technology
- Achieved its designers' objectives –
  - reduced the weight by 40 percent
  - cut acceleration from 0 - 100 kmh to 2.5 seconds from 3.4 seconds
  - increased the power-to-weight ratio, and the car's handling and performance

Lamborghini is the only automaker to have mastered the complete CFRP design-to-production process in-house

Why use carbon composites?

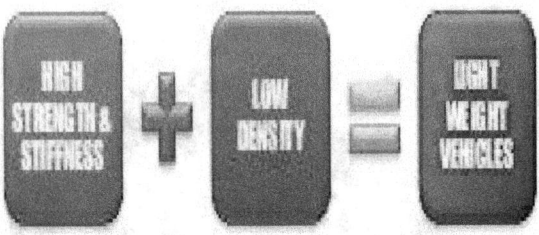

	Carbon Fiber	UD Carbon Composite	Steel	Aluminum
Strength (MPa)	4150	~ 2200	~ 690	~ 415
Modulus (GPa)	245	~ 132	~ 207	~ 69
Density (g/cc)	1.81	~ 1.54	~ 7.8	~ 2.7

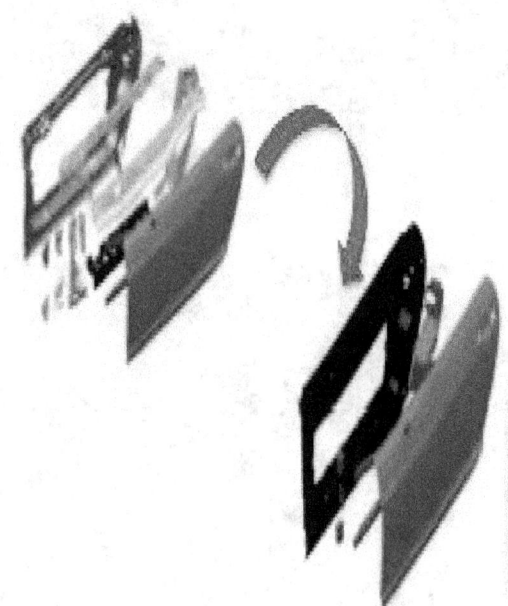

What is needed for broader automotive use of Carbon Fiber Composites?

- Lower cost carbon fiber & intermediate products
- High throughput / low cost manufacturing technologies

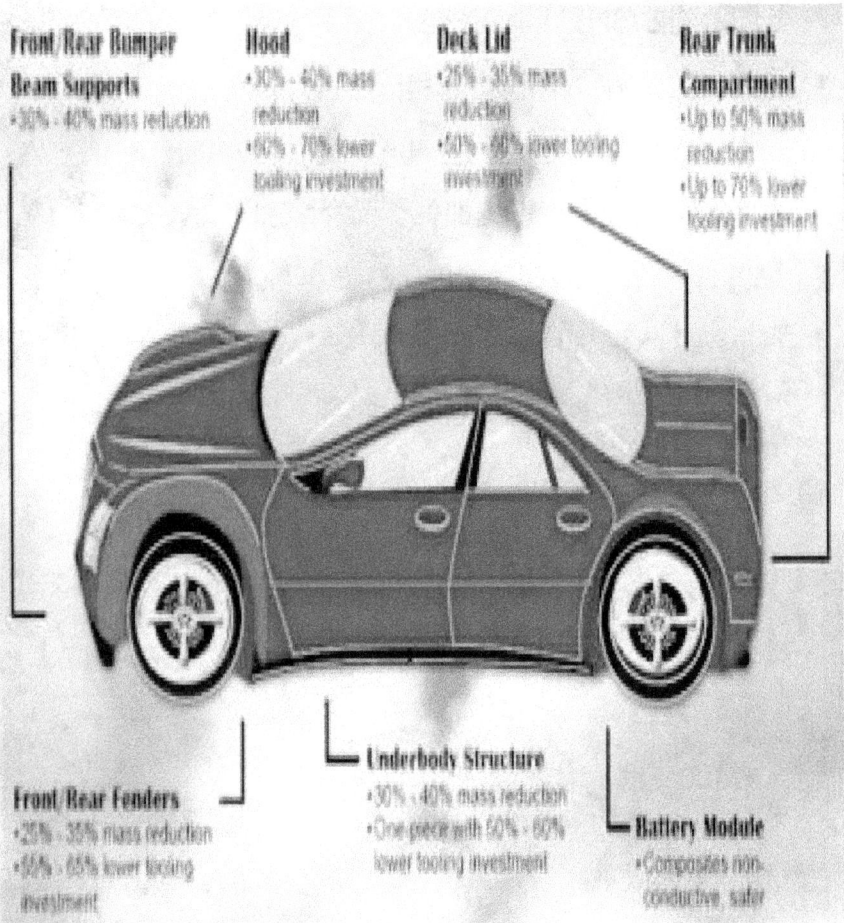

Front/Rear Bumper Beam Supports
- 30% - 40% mass reduction

Hood
- 30% - 40% mass reduction
- 60% - 70% lower tooling investment

Deck Lid
- 25% - 35% mass reduction
- 50% - 60% lower tooling investment

Rear Trunk Compartment
- Up to 50% mass reduction
- Up to 70% lower tooling investment

Front/Rear Fenders
- 25% - 35% mass reduction
- 60% - 65% lower tooling investment

Underbody Structure
- 30% - 40% mass reduction
- One-piece with 50% - 60% lower tooling investment

Battery Module
- Composites non-conductive, safer

TORAY – Carbon Fibre Composite Materials

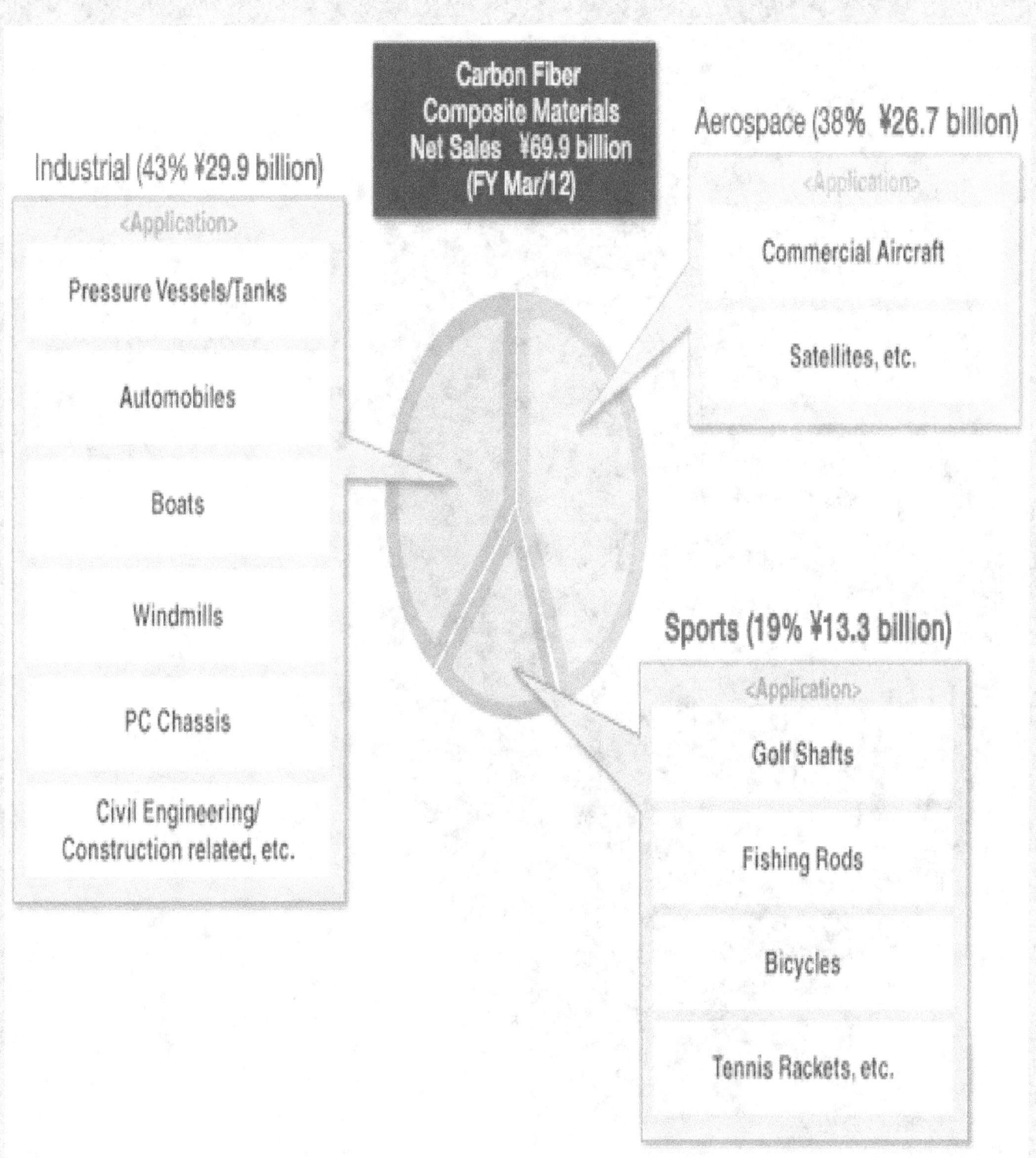

Carbon Fiber Composite Materials Net Sales ¥69.9 billion (FY Mar/12)

Industrial (43% ¥29.9 billion)

<Application>

- Pressure Vessels/Tanks
- Automobiles
- Boats
- Windmills
- PC Chassis
- Civil Engineering/ Construction related, etc.

Aerospace (38% ¥26.7 billion)

<Application>

- Commercial Aircraft
- Satellites, etc.

Sports (19% ¥13.3 billion)

<Application>

- Golf Shafts
- Fishing Rods
- Bicycles
- Tennis Rackets, etc.

Ceramic fibres

✓ Ceramic fibres have as basic components- high purity alumina and silica

✓ The commercial products are both mats and plates

✓ The working temperature is about 1200°C and the melting point of the fibre is about 1750°C.

✓ The density varies from 2.37-3.1g/cc

✓ Excellent high temperature mechanical properties

Properties :

✓ Chemical, micro structural & mechanical stability in air at   high temperatures ~ 1000 °C

✓ Creep resistance (>1200 degree C)

✓ Modulus ~ 200 GPa

✓ % Extn. at break ~ 1%

✓ Strength (room temp.)~ 2 GPa

Processing: flexibility (to be formed as prepegs by weaving) or in filtration by matrix material,Smaller dia (10μm)

(flexibility is reciprocal of 4th power of diameter)

Non-Oxide fibres

✓Silicon Carbide (SiC)

✓Boron Nitride

Silicon Carbide based fibres (SiC)

✓Commercially available as :

• Nicalon (Nippon Carbide)

• Tyranno Lox-M (Ube Industries)

• Sylramic (Dow Corning)

Alumina based fibers

Al_2O_3 99.9 %

Other Oxides (SiO_2, Fe_2O_3, B_2O_3)

Monocrystalline / Polycrystalline

(Commercially available – Altex *Sumitomo* & Nextel - *3M*)

Density ~ 2.7-3.92 g/cc

Young's Modulus~ 152- 414 Gpa

Tensile Strength ~ 1.2- 2.5 Gpa

% Strain at Break ~ 0.3 - 1.12

Stable in oxidizing atmospheres at temp > 1400 ºC

- Reinforcement for CMCs & MMCs

- Ceramic Matrix composites (CMC)

- Metal Matrix composites (MMC)

- Range of oxide & non oxide fibres :

  - dia~10-20μm

  - Large dia~ 100 μm (CVD)

Applications: Gas turbines (aeronautical and ground based), heat exchangers, Candle filter(high temp gas filtration)

Withstand high temp (in air)

GLASS FIBRES

Composition of Glass

	E Glass Range (%)	S Glass Range (%)	C Glass Range (%)
Silicon oxide	52–56	65	64–68
Calcium oxide	16–25	–	11–15
Aluminum oxide	12–16	25	3–5
Boric oxide	5–10	–	4–6
Magnesium oxide	0–5	10	2–4
Sodium oxide and potassium oxide	0–2	–	7–10
Titanium oxide	0–15	–	–
Iron	0–1	–	–
Iron oxide	0–0.8	–	0–0.8
Barium oxide	–	–	0–1

Types and Forms:
Typical 10 – 20 μm diameter

yarn
tow
twisted (Tex of glass yarns 600, 120, 2400)

✓ E-Glass – (Electrical)
• *Good strength, Stiffness, Electrical and weathering properties*

✓ C-Glass (corrosion)
• *Lower strength*

✓ S-Glass (strength)
• *More expensive, High strength, Modulus and temperature*
  resistance

Glass fibres

- **E Glass** -all purpose fibre. Resistant to leaching in water, cheapest
- **C Glass** - acid resistant - not generally used as a reinforcement, chemical corosion resistant
- **A Glass** - typical window glass for comparison purposes andis not used in fibre manufacture
- **S Glass** - highest strength and stiffness, lower density, acid resistant, epoxy resin, aircraft ind.
- **AR glass** - alkali resistant and used for strengtheningcements, rich for zirconoxid, azbestos replacement

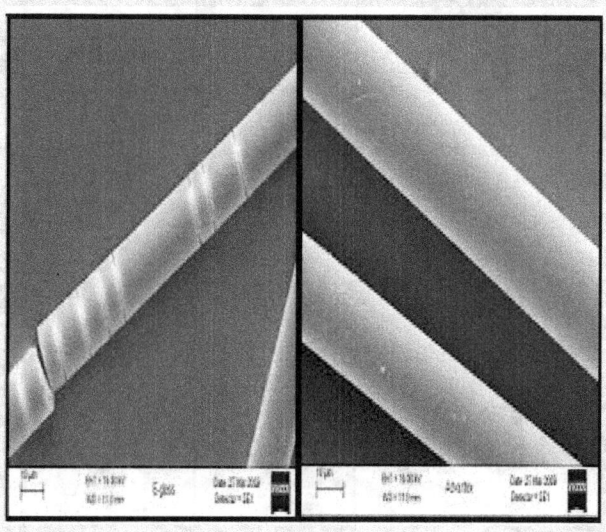
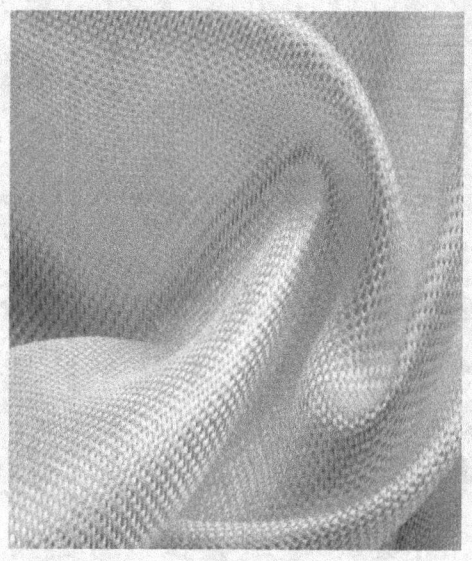

Properties of Glass fibres

Glass Type	ρ (mg/m-3)	σ (GPa)	E (GPa)	α	Tmax (°C)
E - Glass	2.60	3.45	76.0	4.9	550
S – Glass	2.48	4.60	85.5	5.6	650
C – Glass	2.49	3.30	69.0	7.2	600

High strength, stiffness, good tolerance to temperature moisture sensitive and abrasiveness, isotropic.

Advantages

- ✓ Low cost
- ✓ High tensile strength
- ✓ High chemical resistance
- ✓ Excellent insulating properties

Most common reinforcing fibre for polymeric matrices

Disadvantages

- ✓ Low tensile modulus
- ✓ Relatively high specific gravity
- ✓ Sensitivity to abrasion with handling (frequently decreases its tensile strength)
- ✓ Relatively low fatigue resistance
- ✓ High hardness (causes excessive wear on molding dies and cutting tools)

Applications

Major composite applications – moderate strength, low cost
Matrix – unsaturated polyester resin.

- ✓ Leisure boats, Mine Sweepers and High Speed Passenger
- ✓ Ships, Aircraft Radome
- ✓ Interiors Construction – Building Panels, Beams, Gratings, Pipes, Walk ways, cable trays, Storage Tanks, Silos.
- ✓ Electrical Equipment
- ✓ Bathroom interiors and pools
- ✓ Sporting Goods

High density as well abrasive character- a disadvantage

Limitations of glass fibres during use

- ✓ Surface damage (flaws) produced by abrasion either by rubbing against each other or by contact with processing equipment (reduce strength)
- ✓ Strength reduced in presence of water or under sustained loads.
- ✓ Water bleaches out alkali from surface and deepens surface flaws (already present)
- ✓ Tensile strength of glass fibres decreases with increasing time of load duration (surface flaws accelerated under exposure to moisture, under load)

PBO
Fibres

Poly (p-phenylene benzobisoxazole) PBO

✓ Highly fused aromatic heterocyclic polymers for high temp.

   application

✓ High rigidity and form highly ordered structures

✓ Rigid benzoxazole (linear segment)

✓ PBO Commercial~ Zylon (Toyobo).

Propertie
s

✓ Excellent mechanical properties- most thermally stable and flame resistant of all organic fibres

✓ Impressive thermal properties, very high flame resistance and exceptionally high thermal stability (onset of thermal degradation in the 600-700 degree C range)

✓ <u>Tenacity retained (for 3 hrs)</u>

Zylon ~   300 degree C ~65 %

•   400 degree C ~50%

•   500 degree C ~<40%

ZYLON
(PBO)

Chemical resistance

- Exposure to strong acids causes strength losses. However, ZYLON® is more stable than p-Aramid.

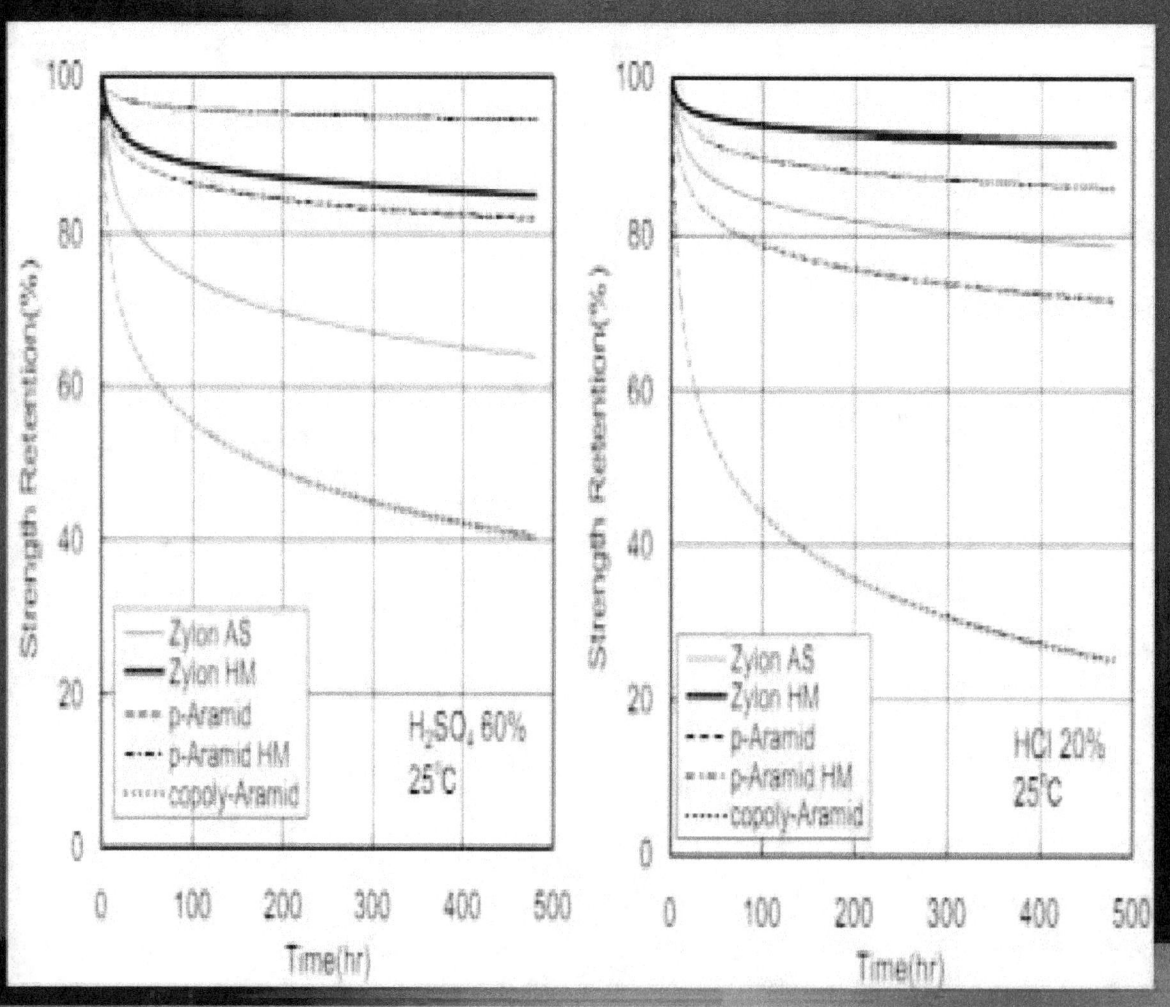

	Tenacity		Modulus		Elonga-tion	Density	Moisture Regain	LOI	Heat Resistance*
	cN/dtex	GPa	cN/dtex	GPa	%	g/cm^3	%		C
Zylon® AS	37	5.8	1150	180	3.5	1.54	2.0	68	650
Zylon® HM	37	5.8	1720	270	2.5	1.56	0.6	68	650
p-Aramid(HM)	19	2.8	850	109	2.4	1.45	4.5	29	550
m-Aramid	4.5	0.65	140	17	22	1.38	4.5	29	400
Steel Fiber	3.5	2.8	290	200	1.4	7.8	0		
HS-PE	35	3.5	1300	110	3.5	0.97	0	16.5	150
PBI	2.7	0.4	45	5.6	30	1.4	15	41	550
Polyester	8	1.1	125	15	25	1.38	0.4	17	260

*Melting or Decomposition Temperature

✓ Good resistance to creep, chemical and abrasion resistance

✓ Poor compressive properties (restrict use in composites)

-failure via kinking

✓ Unlike carbon/inorganic fibres, fibres do not show catastrophic

failure in compression

✓ Kinks develop at yield point in compression

Physical properties of PBO fibres

Density	T. Modulus	T. Strength	Compressive strength
(g/cc)	(Gpa)	(Gpa)	(Gpa)
1.56	280	5.8	0.2

specific strength & modulus ~ highest of all, other materials

✓ **DP is** high - 82-84 %

✓ Rigid rod polymers decompose at high temp without melting and can be dissolved in a few solvents (owing to aromatic structure & rigid back bone)

Conventional melt & solution spinning not feasible

✓ **PBO fibres** ~ Dry-jet-wet spg

✓ Extrusion of polymers solution (liq crystalline phase) under heat and pressure through an air gap into a coagulating bath (~water at Room temp)

✓Followed by washing, drawing and final heat treatment

500-700 $^\circ$ C under tension)

Structure formed- Network of oriented microfibrils

Applications

✓ Excellent thermal and mechanical properties

✓ PBO fabrics are light wt. and flexible, provide improved comfort and mobility and are ideal for heat/flame resistant work wear. e.g. fire fighters

✓ PBO is also ideal for ballistic protection excellent high energy absorption and rapid dissipation of impact by fibrillar morphology.

✓ Heat resistant felts~ conveyor belts (hot fabrication)

✓ Reinforcement~ tyres,belts,cords

✓ Optical fibre cables (good dielectric properties)

✓ Hot gas filters

P O L Y E T H Y L E N E

Polyethylene Fibres : $(-CH_2 - CH_2-)_n$

✓ High strength of C-C bond, small cross-section area.

  (18A°), high orientation, long and regular chain for
  uniformity of structure

✓ Flexible Molecule

✓ Ultra High Mol. Wt.

✓ High Modulus and high strength, low density, high levels of mol. orientation and low cross-section area.

Gel-spun high performance polyethylene fibres types

- ✓ High modulus Polyethylene (HPPE)
- ✓ Extended Chain Polyethylene (ECPE)
- ✓ Ultra strong and high modulus polyethylene

Fibre properties

- ✓ Specific gravity 0.97gm/cc (floats on water) ;Diameter 27μm
- ✓ Tensile Modulus 172 Gpa, Strength 3 Gpa
- ✓ Elongation 3% , High Specific Mod./Strength(10 – 15 times that of good quality steel and to special carbon fibres)
- ✓ Ten fold increase in impact to Kevlar
- ✓ Complete Recovery from deformation
- ✓ Modulus increase with strain rate - low value of Tg
- ✓ Good flexural fatigue resistance (flex life good)
- ✓ Resistance to water ~ do not swell, hydrolyze or degrade
  in water, Hydrophobic

Contd..

- ✓ Highly anisotropic, properties in transverse direction are very poor.

- ✓ Absorb extremely high amounts of energy. Utilized for ballistic protection. Applications in cut resistant gloves, motor helmets, Laminates for balloons (stratospheric exploration) .

- ✓ Improve impact strength of glass / carbon fibres based composites .

- ✓ Good abrasion resistance, low coefficient of friction.

- ✓ Excellent resistance to chemicals (acids and alkalis).

- ✓ Resistance to UV radiation is moderate.

- ✓ No brittle point upto (-269°C), can be used for cryogenic application.

Drawbacks

✓   Melting point ~ 144 to 155 °C
Properties decrease at high
temperature (> 80 to 100 °C)

✓   Fibre is prone to creep – deformation
increases with loading time

Commercial grades

- ✓ **Dyneema**
- ✓ **Spectra** (multifilament yarns)

- ✓ Structure
- • Fibre is highly crystalline > 80 %
- • Extended chain conformation and orientation in direction of fibre.

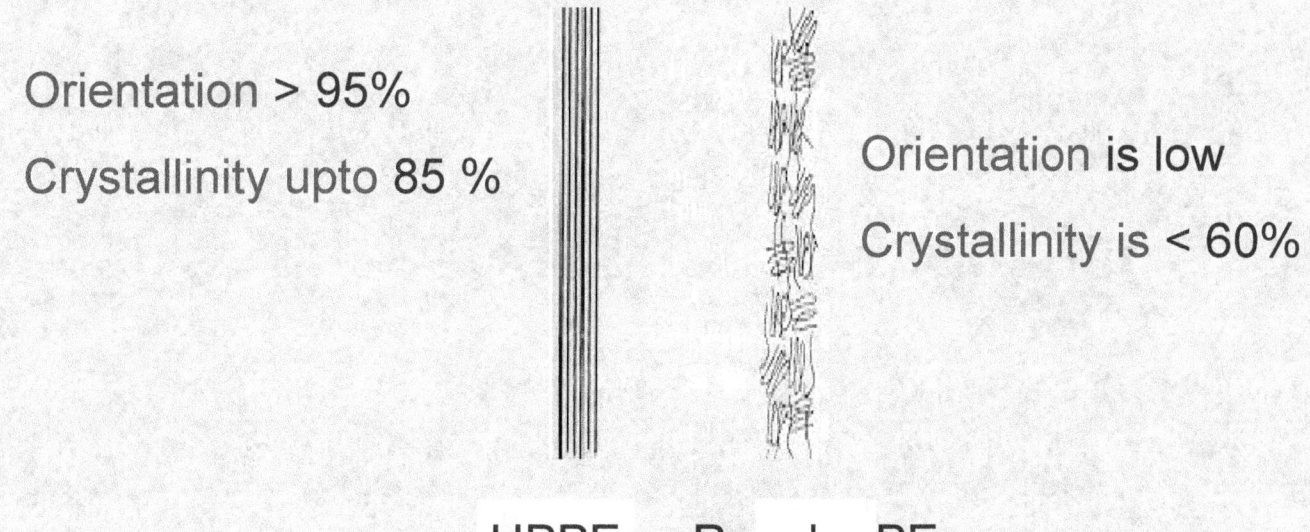

Orientation > 95%

Crystallinity upto 85 %

Orientation is low

Crystallinity is < 60%

HPPE   Regular PE

Manufacturing

✓   Produced from Polyethylene with a very high molecular weight (UHMW – PE)

✓   Spinning from melt is impossible due to high melt viscosity

✓   Drawing is possible only to an extent due to high degree of entanglement.

✓   Solution ~ molecules can disentangle and remain in that state, till fibre is spun

✓   Fibre can be super drawn to get a very high degree of orientation, because it has low degree of entanglement

✓   Process that is based on physical process rather than chemistry

Applications

- ✓ Bullet Proof systems-self reinforcing
- ✓ Hybrid composites with glass and carbon fibres
- ✓ Light weight tiles for structural applications

- Thank You

www.ingramcontent.com/pod-product-compliance
Lightning Source LLC
Chambersburg PA
CBHW081211180526
45170CB00006B/2303